Energy

Copyright © by Harcourt, Inc.

All rights reserved. No part of this publication may be reproduced or transmitted in any form or by any means, electronic or mechanical, including photocopy, recording, or any information storage and retrieval system, without permission in writing from the publisher.

Requests for permission to make copies of any part of the work should be addressed to School Permissions and Copyrights, Harcourt, Inc., 6277 Sea Harbor Drive, Orlando, Florida 32887-6777. Fax: 407-345-2418.

HARCOURT and the Harcourt Logo are trademarks of Harcourt, Inc., registered in the United States of America and/or other jurisdictions.

Printed in the United States of America

ISBN-13: 978-0-15-362085-0

ISBN-10: 0-15-362085-4

1 2 3 4 5 6 7 8 9 10 179 16 15 14 13 12 11 10 09 08 07

Visit *The Learning Site!*
www.harcourtschool.com

Lesson 1

What Are Some Forms of Energy?

VOCABULARY
energy
potential energy
kinetic energy
law of conservation of energy

Energy causes matter to change. Energy from this lightning can split a tree in two or set fire to a forest.

Potential energy is energy due to the position or condition of an object. This skier has high potential energy because he is at the top of a hill.

Kinetic energy is the energy of motion. The skier is changing potential energy to kinetic energy as he moves down the hill.

The **law of conservation of energy** says that the total amount of energy in a system is always the same. Energy can change forms. But it cannot be made or destroyed. The energy in this water will be changed to electrical energy.

READING FOCUS SKILL
MAIN IDEA AND DETAILS
The main idea tells what something is about.
The pieces of information that tell more about the main idea are called the details.

Potential and Kinetic Energy

Energy is the ability to cause change in matter. Energy has different forms. You are using light energy to see this page. You use sound energy when you talk with your friends. You use electrical energy when you play music.

We usually talk about two general types of energy. One type is potential energy. **Potential energy** is energy something has because of its position or condition. Look at the picture below. The skier has high potential energy at the top of the hill. This is because of his position.

As he goes down the hill, his position changes. The potential energy changes into kinetic energy. That's the second kind of energy. **Kinetic energy** is the energy of motion. By the bottom of the hill, almost all of the potential energy has changed into kinetic energy.

At the top of the hill, the skier has a lot of stored or potential energy. As the skier skis down the hill, his potential energy changes to kinetic energy.

Look at the snow at the top of the mountain in the picture below. It has a huge amount of potential energy because it is so high. The snow starts to slide down the mountain. This is called an *avalanche*. The snow's potential energy quickly changes to kinetic energy. Remember that kinetic energy is the energy of motion. So the faster the snow moves downhill, the more kinetic energy it has. It can have so much kinetic energy that it can destroy buildings and trees in its path.

 Explain the difference between kinetic energy and potential energy.

An avalanche on the move. ▶

An avalanche knocked down buildings in this village.

Energy Transformations

One form of energy can be transformed, or changed, into other forms. If you have ever watched a fireworks display, you've watched lots of energy transformations.

The display starts with a rocket. The rocket has stored chemical energy. The chemicals are packed into the rocket in layers. The chemical energy of each layer will be used in order.

Someone lights the rocket with a flame. People near the rocket know to move out of the way. Some of the chemical energy in the rocket changes to thermal energy. People will feel this as heat if they are too close.

The energy in the flame used to light the rocket changes the energy inside the rocket. The potential energy of the chemicals turns into kinetic energy. The rocket moves and flies into the air. More chemical reactions in the rocket make the colored sparks you see. Another reaction releases a bright flash. You see light energy. A bang follows. Some of the chemical energy is changed to sound energy.

 Describe some energy transformations you see during a fireworks display.

▼ Fireworks result from many energy changes.

What energy transformations take place in a toaster oven and in a flashlight?

You also see energy changes when you cook. Suppose you want to make toast. You put a piece of bread into the toaster. The toaster changes electrical energy to thermal energy. This energy toasts the bread. If you look inside the toaster, you will see wires glowing. This is because some of the electrical energy is changed to light energy.

Where does the electrical energy used by the toaster come from? You plug the toaster's wires into an outlet. That outlet is connected to other wires. Those wires go to a generating station where electricity is generated.

Some generating stations use the mechanical energy of moving water to produce electricity. Water held behind a dam has potential energy because of the height of the water. The water flows down through gates in the dam. Its potential energy is changed to mechanical kinetic energy.

Some stations burn a fuel such as oil. Its chemical energy is changed to thermal energy. This energy heats water until it becomes steam. The steam turns large turbines to produce electricity.

Other stations use the potential energy inside an atom (nuclear energy) or the energy of the sun (solar energy).

 Name three ways generating stations get the energy they change to electricity.

Law of Conservation of Energy

The **law of conservation of energy** states that the total amount of energy in a system never changes. It is always the same. Energy cannot be made or destroyed. Energy just changes form.

Think about a lightning storm. Lightning is electrical energy moving between clouds, or between a cloud and the ground. Some of the energy changes to light energy. You see the flash.

▼ Lightning changes to thermal energy when it strikes the ground.

Some energy changes to sound energy. You hear the thunder. The waves of sound energy change to mechanical energy. Your windows might rattle. Your ears may hurt because bones inside them move.

If lightning strikes the Earth, some of its electrical energy changes into thermal energy. This energy is transferred as heat. This heat can melt sand on a beach, split a tree in two, or start a forest fire.

No energy is lost during a lightning storm. No new energy is made. Energy just changes form.

 Explain how a lightning storm shows the law of conservation of energy.

Complete this main idea statement.

1. _____ causes matter to change.

Complete these detail statements.

2. The _____ energy of an object depends on its position or condition.

3. _____ energy is the energy of motion.

4. The law of conservation of _____ states that energy cannot be made or destroyed, but it can change form.

9

Lesson 2

What Are Waves?

VOCABULARY
wave
wavelength
amplitude
frequency
electromagnetic spectrum

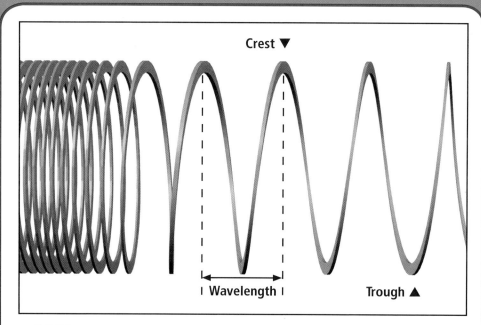

A **wave** carries energy through matter or space. A wave has a crest at the top and a trough at the bottom. The distance from the middle of one crest to the middle of the next is called **wavelength**.

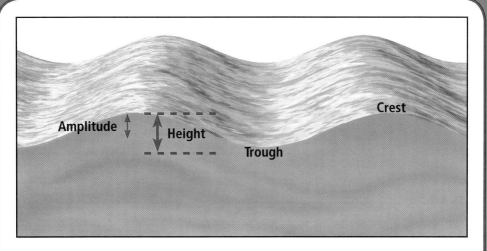

The **amplitude** of a wave tells how far the wave moves from its resting position. It's the distance from the resting position to the top of the crest or to the bottom of the trough.

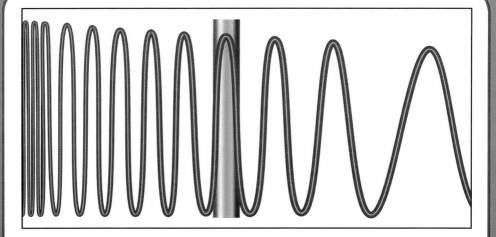

Light and other waves like it make up the **electromagnetic spectrum.** Visible light is in the middle of this diagram. Waves to the left have short wavelengths and high frequencies. Waves to the right have long wavelengths and low frequencies. The **frequency** is the number of waves in a given amount of time.

READING FOCUS SKILL

The main idea is the most important idea in a text or story. The details give more information about the main idea.

Look for what a wave is and then look for details about waves.

Waves

Have you ever sat on the beach and just watched the waves come in? Each wave carries water to the beach. If you get into the water, you'll see how much energy a wave carries. It can knock you down! "Wave" has a scientific meaning too. A **wave** is a movement that travels through matter or space and carries energy.

The kind of wave that travels through matter is called a *mechanical wave.* An earthquake is a mechanical wave. You'll read about sound waves later. These are mechanical waves. Light waves are not. They do not need matter. They travel through space.

A wave has a high part and a low part. The high part is called the *crest*. The low part is called a *trough*. We also call the top of a mountain the crest. We sometimes call a gutter a trough.

Sometimes we want to know how long a wave is. We can measure from the middle of the crest of one wave to the middle of the crest of another wave. Or we can measure from the middle of the trough of one wave to the middle of the trough of another. We would get the same measurement. This measurement is called a **wavelength**.

Suppose you draw a line in the middle of a wave. The line is halfway between the crest and the trough. This is a wave's "resting position." We sometimes want to know how far a wave moves from its resting position. This is called amplitude. A wave's **amplitude** is the distance from the resting position to the top of the crest or the bottom of the trough.

 Use the words *crest, trough, wavelength, amplitude,* and *resting position* to describe a wave.

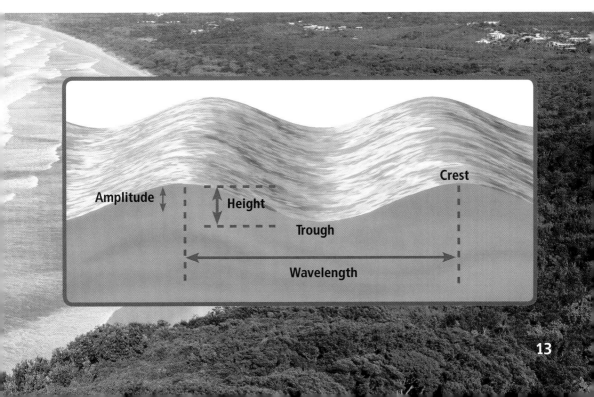

Sound Waves

Think about how a guitar makes sound. You pluck the string and make it vibrate. The string pushes on the air next to it. This push compresses, or squeezes, air particles together. This is called compression.

A sound wave has two parts. One part is the compression part. This is where particles are pushed together. The other part is the rarefaction (rair•uh•FAK•shuhn) part. In the rarefaction part, the particles are far apart. The picture below shows sound waves. It shows two compressions. A single sound wave is made of one compression and one rarefaction. How many sound waves do you see in the picture?

Now, think about the guitar string again. If you look closely at a guitar string after you pluck it, you'll see it keeps vibrating. The number of times it vibrates in a given amount of time is its **frequency**. Frequency is measured in hertz. One hertz equals one vibration per second.

 Explain how a sound wave travels through air.

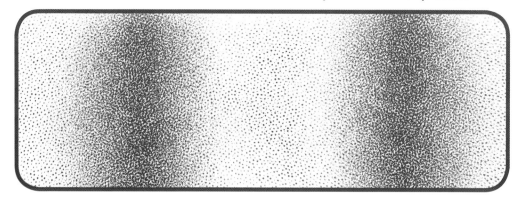

▼ A sound wave is called a compression wave. Find where the particles are compressed.

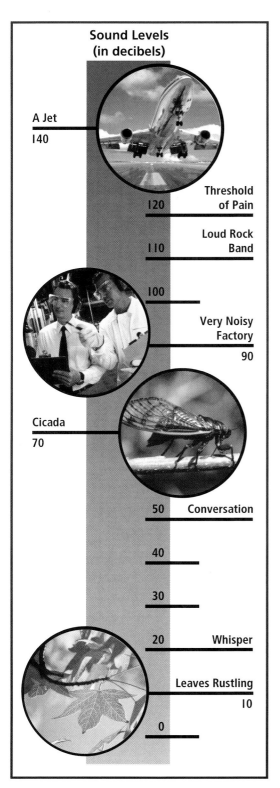

Sound Levels (in decibels)

- A Jet — 140
- 120 Threshold of Pain
- 110 Loud Rock Band
- 100
- 90 Very Noisy Factory
- Cicada — 70
- 50 Conversation
- 40
- 30
- 20 Whisper
- 10 Leaves Rustling
- 0

◀ **Louder sounds have greater amplitude.**

One way to describe sound is by its pitch. The *pitch* of a sound is how low or high a sound is. The pitch is determined by the frequency. The higher the frequency, the higher the pitch.

Another way to describe sound is by its loudness. If you strum a guitar gently, the sound is soft. If you strum it with more force, the vibration has more energy. The sound is louder. Loudness is measured in *decibels*. The sound of leaves rustling on a tree has a loudness of 10 decibels. Loud sounds over 80 decibels can, over time, damage your hearing. You have to be careful to protect your ears from constant loud noises.

 Tell how loudness is related to energy.

Light and Electromagnetic Waves

Light is a form of energy that travels in waves. But light waves don't need matter. They can travel through space.

Light and some other waves are electromagnetic waves. This means they are vibrating electric and magnetic fields that carry energy. These kinds of waves make up the **electromagnetic spectrum**. Visible light is part of this spectrum. You can see it in the middle of the diagram below.

The waves to the left of visible light have shorter wavelengths and higher frequencies. They have less energy than visible light. They include infrared waves, microwaves, and radio waves. Infrared waves are heat waves. They are made by things that are hot.

The waves to the right of visible light have longer wavelengths and lower frequencies. They include X rays and ultraviolet waves. Ultraviolet (UV) rays cause sunburns. They can also harm your eyes and skin. It is important to protect yourself from UV rays.

 What makes up the electromagnetic spectrum?

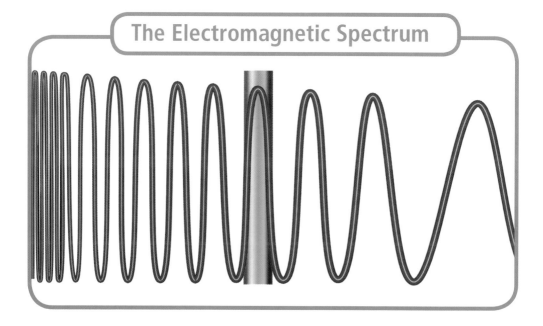

The Electromagnetic Spectrum

Light Energy from the Sun

The sun provides energy for life on Earth. Energy from the sun warms us. It also warms Earth's air, land, and water.

The sun warms the Earth's land unevenly. This causes the winds that move weather systems. Energy from the sun warms water and causes it to evaporate. This is one step in the water cycle. The water cycle makes life on Earth possible.

Energy from the sun also provides us with all of our food. Plants take in sunlight and make food. We eat the plants or the animals that eat the plants. All the energy we get comes fom the sun.

The picture on the left shows how the sun normally looks to us. The picture above shows the sun as seen in ultraviolet light. The picture below shows an X-ray view of the sun.

Review

Complete these main idea statements.

1. The top part of a wave is called the _____ . The bottom part of the wave is called the _____.
2. _____ is the number of vibrations per second.

Complete these detail statements.

3. Sound waves are made of a _____ and a rarefaction.
4. _____ _____ are an example of heat waves.

Lesson 3

How Does Light Behave?

VOCABULARY
reflection
refraction
diffraction
transparent
translucent
opaque

Reflection is the bouncing of light off a surface. This lighthouse has a mirror that reflects light out to the ships.

Refraction is the bending of light as it goes from the surface of one material to another.

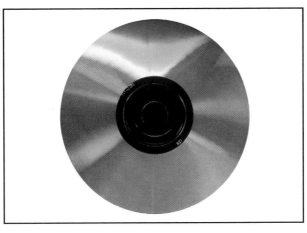

Diffraction is the bending of light around the edge of an object. Diffraction produces a rainbow effect.

A **transparent** material lets most of the light through it. Clear glass is transparent. A **translucent** material lets some of the light through it. Wax paper is translucent. An **opaque** material does not let light pass through it. Wood is opaque.

READING FOCUS SKILL

CAUSE AND EFFECT

The cause is what makes something happen. The effect is what happens as a result. Look for the effects of light on different objects.

How Light Behaves

Look at the beam of light from the lighthouse. Notice how the light travels in a straight line. Light travels in a straight line until it hits something.

When light hits an object, some of it bounces off the surface. Light bouncing off a surface is called **reflection**. A smooth surface like a mirror reflects light well. This is why you can see your image in a mirror.

The amount of light an object reflects depends on its surface. A smooth surface like a mirror reflects almost all the light that strikes it. A rough surface, like that of soil, reflects much less light. It absorbs the rest of the light.

 Explain what causes reflection.

A prism separates white light into the colors of the spectrum.

The light above is sunlight. You can see the colors in it because it passed through a prism. Sunlight appears white to us. But it's really a mixture of different wavelengths of light. Each wavelength is a different color. We see all the wavelengths together. That's why we see sunlight as white. But if a prism like the one above separates the light, we can see the different colors that make up sunlight. We see each wavelength as a different color.

Reflection and absorption explain why we see colors. When we look at a green object, we see the color green because the object reflects green light. It absorbs all the other colors.

A white object reflects all wavelengths of light. A black object absorbs all the wavelengths. It reflects very little light. We don't see any colors.

 Tell why you see a strawberry as red and the leaves as green.

21

Refraction and Diffraction

Have you ever looked at a straw in a glass of water? The straw looks like it bends. But it doesn't. It looks bent because of refraction. **Refraction** is the bending of light.

The speed of light is different through different materials. Light travels more slowly through dense materials. So light travels more slowly through water than it does through air. When the light leaves water and enters the air, it speeds up. Light from the part of the straw above water reaches your eye on one path. Light from the part of the straw below water reaches your eye on a different path. You see the straw in two parts. It looks bent.

Light also bends slightly when it passes the edge of an object. This is why the edge of a shadow looks a little blurry. **Diffraction** is the bending of light around the edge of an object.

We can see the effects of diffraction when we see the different colors at sunset. Sunlight bends around the particles in the atmosphere. Longer wavelengths are affected more than shorter ones are. Red light has a longer wavelength. So we see red colors at sunset.

 What causes diffraction?

Light is diffracted by particles in the atmosphere during a sunset.

22

Transparent, Translucent, and Opaque

We can sort objects into three groups by how they affect light. An object is **transparent** if it lets most of the light through it. Clear objects like windows are transparent.

An object is **translucent** if it lets only some light through it. Frosted glass is translucent.

Something is **opaque** if it doesn't let any light pass through it. Light striking an opaque object is either reflected or absorbed. You and your friends are all opaque.

 Explain the difference between transparent, translucent, and opaque.

This vase is translucent. ▶

Review

 Complete these cause-and-effect statements.

1. Light striking and bouncing off an object causes a _____.
2. The effect of light striking the edge of an object is called _____.
3. The flower stem in a vase appears bent due to _____.
4. If an object is _____, you can see through it clearly.

GLOSSARY

amplitude (AM•pluh•tood) The distance in a wave from the resting position to the top of the crest or the bottom of the trough.

diffraction (dih•FRAK•shuhn) The bending of light around the edge of an object.

electromagnetic spectrum (ee•lek•troh•mag•NET•ik SPEK•truhm) All energy waves that travel at the speed of light in a vacuum; includes radio, infrared, visible, ultraviolet, x rays, and gamma rays.

energy (EN•er•jee) The ability to cause change in matter.

frequency (FREE•kwuhn•see) The number of vibrations or waves in a given amount of time.

kinetic energy (kih•NET•ik EN•er•jee) The energy of motion.

law of conservation of energy (LAW uv kahn•ser•VAY•shuhn uv EN•er•jee) The rule stating that the total amount of energy in a closed system is always the same—energy cannot be created or destroyed.

opaque (oh•PAYK) Not allowing any light to pass through.

potential energy (poh•TEN•shuhl EN•er•jee) Energy that is due to the position or condition of an object.

reflection (rih•FLEK•shuhn) The bouncing of light off a surface.

refraction (rih•FRAK•shuhn) The bending of light as it passes from the surface of one material to another.

translucent (tranz•LOO•suhnt) Allowing some light to pass through.

transparent (tranz•PAR•uhnt) Allowing almost all light to pass through.

wave (WAYV) A disturbance that carries energy through matter or space.

wavelength (WAYV•length) The distance from the middle of the crest of one wave to the middle of the next crest.